Acres of Diamond

Russell Herman Conwell

and

The Way to Wealth

Benjamin Franklin

Two Books in One

Acres of Diamond

Russell Herman Conwell

and

The Way to Wealth

Benjamin Franklin

Two Books in One

IAP © 2009

Printed in the United States of America.

Franklin, Benjamin and Conwell, Russell Herman.

Acres of Diamond. The Way to Wealth. Two Books in One / Herman Russell Conwell Benjamin Franklin. - 1st ed.

1. History.

Table of Contents

Introduction

Benjamin Franklin (1706-1790) was an author, printer, political theorist, politician, scientist, inventor, statesman, diplomat and one of the Founding Fathers of the United States. Franklin and George Washington stand together as the greatest heroes of the American Revolution. Later Benjamin Franklin became very wealthy and wrote *The Way to Wealth*, which is a collection of adages and advice presented in Poor Richard's Almanac. Many of the adages found in his book continue to be familiar today. Franklin also played a very important role in the establishment of the University of Pennsylvania.

Russell Herman Conwell (1843 –1925) was an American Baptist minister, educator, lawyer, writer, and orator. He founded the Temple University, in Philadelphia, Pennsylvania, and he wrote *Acres of Diamonds*, which originated as a speech he delivered over 6,000 times worldwide. During the American Civil War, he served in the Union Army from 1862 to 1864. According to his message, success in the United States not only was possible, it was a moral obligation. He said "you have no right to be poor. It is your duty to be rich."

8

Franklin about the Poor Richard's Almanac in his autobiography

In 1732 I first published my Almanac under the name of *Richard Saunders*; it was continued by me about twenty-five years, and commonly called *Poor Richard's Almanac*. I endeavored to make it both entertaining and useful, and it accordingly came to be in such demand, that I reaped considerable profit from it, vending annually near ten thousand. And observing that it was generally read, (scarce any neighborhood in the province being without it,) I considered it as a proper vehicle for conveying instruction among the common people, who bought scarcely any other books. I therefore filled all the little spaces, that occurred between the remarkable days in the Calendar, with proverbial sentences, chiefly such as inculcated industry and frugality, as the means of procuring wealth, and thereby securing virtue; it being more difficult for a man in want to act always honestly, as (to use here one of those proverbs) It is hard for an empty sack to stand upright.

These proverbs, which contained the wisdom of many ages and nations, I assembled and formed into a connected discourse prefixed to the almanac of 1757 as the harangue of a wise old man to the people attending an auction. The bringing all these scattered counsels thus into a focus enabled them to make greater impression. The piece, being universally approved, was copied in all the newspapers of the American continent, reprinted in Britain on a large sheet of paper to be stuck up in houses; two translations were made of it in France, and great numbers bought by the clergy and gentry to distribute gratis among their poor parishioners and tenants. In Pennsylvania, as it discouraged useless expense in foreign superfluities, some thought it had its share of influence in producing that growing plenty of money which was observable for several years after its publication.

The Way to Wealth

Benjamin Franklin

Courteous Reader,

I have heard that nothing gives an author so great pleasure, as to find his works respectfully quoted by other learned authors. This pleasure I have seldom enjoyed; for though I have been, if I may say it without vanity, an eminent author of almanacs annually now a full quarter of a century, my brother authors in the same way, for what reason I know not, have ever been very sparing in their applauses; and no other author has taken the least notice of me, so that did not my writings produce me some solid pudding, the great deficiency of praise would have quite discouraged me.

I concluded at length, that the people were the best judges of my merit; for they buy my works; and besides, in my rambles, where I am not personally known, I have frequently heard one or other of my adages repeated, with, as Poor Richard says, at the end on't; this gave me some satisfaction, as it showed not only that my instructions were regarded, but discovered likewise some respect for my authority; and I own, that to encourage the practice of remembering and repeating those wise sentences, I have sometimes quoted myself with great gravity.

Judge then how much I must have been gratified by an incident I am going to relate to you. I stopped my horse lately where a great number of people were collected at a vendue of merchant goods. The hour of sale not being come, they were conversing on the badness of the times, and one of the company called to a plain clean old man, with white locks, "Pray, Father Abraham, what think you of the times? Won't these heavy taxes quite ruin the country? How shall we be ever able to pay them? What would you advise us to?" Father Abraham stood up, and replied, "If you'd have my advice, I'll give it you in short, for a *word to the wise is enough*, and *many words won't fill a bushel*, as *Poor Richard says*." They joined in desiring him to speak his mind, and gathering round him, he proceeded as follows:

"Friends, says he, and neighbors, the taxes are indeed very heavy, and if those laid on by the government were the only ones we had to pay, we might more easily discharge them; but we have many others, and much more grievous to some of us. We are taxed twice as much by our idleness, three times as much by our pride, and four times as much by our folly, and from these taxes the commissioners cannot ease or deliver us by allowing an abatement. However let us hearken to good advice, and something may be done for us; *God helps them that help themselves*, as Poor Richard says, in his almanac of 1733.

"It would be thought a hard government that should tax its people one tenth part of their time, to be employed in its service. But idleness taxes many of us much more, if we reckon all that is spent in absolute sloth, or doing of nothing, with that which is spent in idle employments or amusements, that amount to nothing. Sloth, by bringing on diseases, absolutely shortens life. *Sloth, like rust, consumes faster than labor wears, while the used key is always bright,* as Poor Richard says. But *dost thou love life, then do not squander time, for that's the stuff life is made of,* as Poor Richard says. How much more than is necessary do we spend in sleep! forgetting that *the sleeping fox catches no poultry,* and that *there will be sleeping enough in the grave,* as Poor Richard says. If time be of all things the most precious, *wasting time* must be, as Poor Richard says, *the greatest prodigality,* since, as he elsewhere tells us, *lost time is never found again,* and what we call *time-enough, always proves little enough*: let us then be up and be doing, and doing to the purpose; so by diligence shall we do more with less perplexity. *Sloth makes all things difficult, but industry all easy,* as Poor Richard says; and *he that rises late, must trot all day, and shall scarce overtake his business at night.* While *laziness travels so slowly, that poverty soon overtakes him,* as we read in Poor Richard, who adds, *drive thy business, let not that drive thee*; and *early to bed, and early to rise, makes a man healthy, wealthy and wise.*

"So what signifies wishing and hoping for better times. We may make these times better if we bestir ourselves. *Industry need not wish,* as Poor Richard says, and *he that lives upon hope will die fasting. There are no gains, without pains,* then *help hands, for I have no lands*, or if I have, they are smartly taxed. And, as Poor Richard likewise observes, *he that hath a trade hath an estate*, and *he that hath a calling hath an office of profit and honor*; but then the trade must be worked at, and the calling well followed, or neither the estate, nor the office, will enable us to pay our taxes. If we are industrious we shall never starve; for, as Poor Richard says, *at the working man's house hunger looks in, but dares not enter*. Nor will the bailiff nor the constable enter, *for industry pays debts, while despair increases them*, says Poor Richard. What though you have found no treasure, nor has any rich relation left you a legacy, *diligence is the mother of good luck*, as Poor Richard says, and *God gives all things to industry*. Then *plough deep, while sluggards sleep, and you shall have corn to sell and to keep*, says Poor Dick. Work while it is called today, for you know not how much you may be hindered tomorrow, which makes Poor Richard say, *one today is worth two tomorrows*; and farther, *have you somewhat to do tomorrow, do it today*. If you were a servant, would you not be ashamed that a good master should catch you idle? Are you then your own master, *be ashamed to catch yourself idle*, as Poor Dick says. When there is so much to be done for yourself, your family, your country, and your gracious king, be up by peep of day; *let not the sun look down and say, inglorious here he lies*. Handle your tools without mittens; remember that *the cat in gloves catches no mice*, as Poor Richard says. 'Tis true there is much to be done, and perhaps you are weak handed, but stick to it steadily, and you will

16

see great effects, for *constant dropping wears away stones*, and by *diligence and patience the mouse ate in two the cable; and little strokes fell great oaks*, as Poor Richard says in his almanac, the year I cannot just now remember.

"Methinks I hear some of you say, must a man afford himself no leisure? I will tell thee, my friend, what Poor Richard says, *employ thy time well if thou meanest to gain leisure; and, since thou art not sure of a minute, throw not away an hour.* Leisure is time for doing something useful; this leisure the diligent man will obtain, but the lazy man never; so that, as Poor Richard says, *a life of leisure and a life of laziness are two things.* Do you imagine that sloth will afford you more comfort than labor? No, for as Poor Richard says, *trouble springs from idleness, and grievous toil from needless ease. Many without labor would live by their wits only, but they break for want of stock.* Whereas industry gives comfort, and plenty, and respect: *fly pleasures, and they'll follow you. The diligent spinner has a large shift,* and *now I have a sheep and a cow, everybody bids me good morrow,* all which is well said by Poor Richard.

"But with our industry, we must likewise be steady, settled and careful, and oversee our own affairs with our own eyes, and not trust too much to others; for, as Poor Richard says,

I never saw an oft removed tree,
Nor yet an oft removed family,

That throve so well as those that settled be.

"And again, *three removes is as bad as a fire,* and again, *keep the shop, and thy shop will keep thee;* and again, *if you would have your business done, go; if not, send.* And again,
He that by the plough would thrive,
Himself must either hold or drive.

"And again, *the eye of a master will do more work than both his hands;* and again, *want of care does us more damage than want of knowledge;* and again, *not to oversee workmen is to leave them your purse open.* Trusting too much to others' care is the ruin of many; for, as the almanac says, *in the affairs of this world men are saved not by faith, but by the want of it;* but a man's own care is profitable; for, says Poor Dick, *learning is to the studious,* and *riches to the careful,* as well as *power to the bold,* and *Heaven to the virtuous.* And farther, *if you would have a faithful servant, and one that you like, serve yourself.* And again, he advises to circumspection and care, even in the smallest matters, because sometimes *a little neglect may breed great mischief;* adding, *for want of a nail the shoe was lost; for want of a shoe the horse was lost, and for want of a horse the rider was lost,* being overtaken and slain by the enemy, all for want of care about a horse-shoe nail.

"So much for industry, my friends, and attention to one's own business; but to these we must add frugality, if we would make our industry more certainly successful. A man may, if he knows not how to save as he gets, *keep his nose all his life to the grindstone,* and die not worth a *groat[1]* at last. *A fat kitchen makes a lean will,* as Poor Richard says; and,

Many estates are spent in the getting,
Since women for tea forsook spinning and knitting,
And men for punch forsook hewing and splitting.

If you would be wealthy, says he, in another almanac, *think of saving as well as of getting: the Indies have not made Spain rich, because her outgoes are greater than her incomes.* Away then with your expensive follies, and you will not have so much cause to complain of hard times, heavy taxes, and chargeable families; for, as Poor Dick says,

Women and wine, game and deceit,
Make the wealth small, and the wants great.

And farther, *what maintains one vice, would bring up two children.* You may think perhaps that a little tea, or a little punch now and then, diet a little more costly, clothes a little finer, and a little entertainment now and then, can be no great Matter; but remember what Poor Richard says, *few people make much,* and farther, *beware of little expenses; a small leak will sink a great ship,*

---

[1] A former English silver coin worth four pence, used until the 17th century

and again, *who dainties love, shall beggars prove,* and moreover, *fools make Feasts, and wise men eat them.*

"Here you are all got together at this vendue of fineries and knick-knacks. You call them goods, but if you do not take care, they will prove evils to some of you. You expect they will be sold cheap, and perhaps they may for less than they cost; but if you have no occasion for them, they must be dear to you. Remember what Poor Richard says, *buy what thou hast no need of, and ere long thou shalt sell thy necessaries.* And again, *at a great pennyworth pause a while*: he means, that perhaps the cheapness is apparent only, and not real; or the bargain, by straitening thee in thy business, may do thee more harm than good. For in another place he says, *many have been ruined by buying good pennyworths.* Again, Poor Richard says, *'tis foolish to lay our money in a purchase of repentance*; and yet this folly is practiced every day at vendues, for want of minding the almanac. *Wise men*, as Poor Dick says, *learn by others' harms, fools scarcely by their own*, but, *felix quem faciunt aliena pericula cautum.* Many a one, for the sake of finery on the back, have gone with a hungry belly, and half starved their families; *silks and satins, scarlet and velvets*, as Poor Richard says, *put out the kitchen fire.* These are not the necessaries of life; they can scarcely be called the conveniences, and yet only because they look pretty, how many want to have them. The artificial wants of mankind thus become more numerous than the natural; and, as Poor Dick says, *for one* poor *person, there are an hundred* indigent. By these, and other extravagancies, the genteel are reduced to poverty, and forced to borrow of those whom they formerly despised, but who through industry and frugality have maintained their standing; in which case it appears plainly, that a *ploughman on his legs is higher than a gentleman on his knees,* as Poor Richard says. Perhaps they have had a small estate left them, which they knew not the getting of; they think *'tis day, and will never be night;* that a little to be spent out of so much, is not worth minding; *(a child and a fool,* as Poor Richard says, *imagine twenty shillings and twenty years can never be spent)* but, *always taking out of the meal-tub, and never putting in, soon comes to the bottom;* then, as Poor Dick says, *when the well's dry, they know the worth of water.* But this they might have known before, if they had taken his advice; *if you would know the value of money, go and try to borrow some,* for, *he that goes a borrowing goes a sorrowing,* and indeed so does he that lends to such people, when he goes to get it in again. Poor Dick farther advises, and says,

Fond pride of dress, is sure a very curse;
E'er fancy you consult, consult your purse.

And again, *pride is as loud a beggar as want, and a great deal more saucy.* When you have bought one fine thing you must buy ten more, that your appearance maybe all of a piece; but Poor Dick says, *'tis easier to suppress the first desire than to satisfy all that follow*

21

it. And 'tis as truly folly for the poor to ape the rich, as for the frog to swell, in order to equal the ox.

Great estates may venture more,
But little boats should keep near shore.

'Tis however a folly soon punished; for *pride that dines on vanity sups on contempt*, as Poor Richard says. And in another place, *pride breakfasted with plenty, dined with poverty, and supped with infamy*. And after all, of what use is this *pride of appearance*, for which so much is risked, so much is suffered? It cannot promote health; or ease pain; it makes no increase of merit in the person, it creates envy, it hastens misfortune.

What is a butterfly? At best
He's but a caterpillar dressed.
The gaudy fop's his picture just,

as Poor Richard says.

"But what madness must it be to run in debt for these superfluities! We are offered, by the terms of this vendue, six months' credit; and that perhaps has induced some of us to attend it, because we cannot spare the ready money, and hope now to be fine without it. But, ah, think what you do when you run in debt; *you give to another power over your liberty*. If you cannot pay at the time, you will be ashamed to see your creditor; you will be in fear when you speak to him, you will make poor pitiful sneaking excuses, and by degrees come to lose you veracity, and sink into base downright lying; for, as Poor Richard says, *the second vice is lying, the first is running in debt*. And again to the same purpose, *lying rides upon debt's back*. Whereas a freeborn Englishman ought not to be ashamed or afraid to see or speak to any man living. But poverty often deprives a man of all spirit and virtue: *'tis hard for an empty bag to stand upright*, as Poor Richard truly says. What would you think of that Prince, or that government, who should issue an edict forbidding you to dress like a gentleman or a gentlewoman, on pain of imprisonment or servitude? Would you not say, that you are free, have a right to dress as you please, and that such an edict would be a breach of your privileges, and such a government tyrannical? And yet you are about to put yourself under that tyranny when you run in debt for such dress! Your creditor has authority at his pleasure to deprive you of your liberty, by confining you in goal for life, or to sell you for a servant, if you should not be able to pay him! When you have got your bargain, you may, perhaps, think little of payment; but *creditors*, Poor Richard tells us, *have better memories than debtors*, and in another place says, *creditors are a superstitious sect, great observers of set days and times*. The day comes round before you are aware, and the demand is made before you are prepared to satisfy it. Or if you bear your debt in mind, the term which at first seemed so long, will, as it lessens, appear extremely short. Time will seem to have added wings to his heels as well as shoulders. *Those have a short Lent*, says Poor Richard, *who owe money to be paid at Easter*. Then since, as he says, *the borrower is a slave to the lender, and the debtor to the creditor*, disdain the chain, preserve your freedom; and maintain your independency: be industrious and free; be frugal and free. At present, perhaps, you may think yourself in thriving circumstances, and that you can bear a little extravagance without injury; but,

For age and want, save while you may;
No morning sun lasts a whole day,

as Poor Richard says. Gain may be temporary and uncertain, but ever while you live, expense is constant and certain; and *'tis easier to build two chimneys than to keep one in fuel*, as Poor Richard says. So *rather go to bed without dinner than rise in debt*.

Get what you can, and what you get hold;
'Tis the stone that will turn all your lead into gold,

as Poor Richard says. And when you have got the philosopher's stone, sure you will no longer complain of bad times, or the difficulty of paying taxes.

"This doctrine, my friends, is reason and wisdom; but after all, do not depend too much upon your own industry, and frugality, and prudence, though excellent things, for they may all be blasted without the blessing of heaven; and therefore ask that blessing humbly, and be not uncharitable to those that at present seem to want it, but comfort and help them. Remember Job suffered, and was afterwards prosperous.

"And now to conclude, *experience keeps a dear school, but fools will learn in no other, and scarce in that,* for it is true, *we may give advice, but we cannot give conduct,* as Poor Richard says: however, remember this, *they that won't be counseled, can't be helped,* as Poor Richard says: and farther, that *if you will not hear reason, she'll surely rap your knuckles.*"

Thus the old gentleman ended his harangue. The people heard it, and approved the doctrine, and immediately practiced the contrary, just as if it had been a common sermon; for the vendue opened, and they began to buy extravagantly, notwithstanding all his cautions, and their own fear of taxes. I found the good man had thoroughly studied my almanacs, and digested all I had dropped on those topics during the course of five-and-twenty years. The frequent mention he made of me must have tired any one else, but my vanity was wonderfully delighted with it, though I was conscious that not a tenth part of the wisdom was my own which he ascribed to me, but rather the gleanings I had made of the sense of all ages and nations. However, I resolved to be the better for the echo of it; and though I had at first determined to buy stuff for a new coat, I went away resolved to wear my old one a little longer. Reader, if thou wilt do the same, thy profit will be as great as mine. I am, as ever, thine to serve thee,

Richard Saunders.

July 7, 1757.

Acres of Diamond

Russell Herman Conwell

AN APPRECIATION

THOUGH Russell H. Conwell's Acres of Diamonds have been spread all over the United States, time and care have made them more valuable, and now that they have been reset in black and white by their discoverer, they are to be laid in the hands of a multitude for their enrichment.

In the same case with these gems there is a fascinating story of the Master Jeweler's life-work which splendidly illustrates the ultimate unit of power by showing what one man can do in one day and what one life is worth to the world.

As his neighbor and intimate friend in Philadelphia for thirty years, I am free to say that Russell H. Conwell's tall, manly figure stands out in the state of Pennsylvania as its first citizen and ``The Big Brother'' of its seven millions of people.

From the beginning of his career he has been a credible witness in the Court of Public Works to the truth of the strong language of the New Testament Parable where it says, ``If ye have faith as a grain of mustard-seed, ye shall say unto this mountain, `Remove hence to yonder place,' AND IT SHALL REMOVE AND NOTHING SHALL BE IMPOSSIBLE UNTO YOU.

As a student, schoolmaster, lawyer, preacher, organizer, thinker and writer, lecturer, educator, diplomat, and leader of men, he has made his mark on his city and state and the times in which he has lived. A man dies, but his good work lives.

His ideas, ideals, and enthusiasms have inspired tens of thousands of lives. A book full of the energy from a master workman is just what every young man cares for.

1915. *John Wanamaker*

ACRES OF DIAMONDS

Friends.--This lecture has been delivered under these circumstances: I visit a town or city, and try to arrive there early enough to see the postmaster, the barber, the keeper of the hotel, the principal of the schools, and the ministers of some of the churches, and then go into some of the factories and stores, and talk with the people, and get into sympathy with the local conditions of that town or city and see what has been their history, what opportunities they had, and what they had failed to do-- and every town fails to do something--and then go to the lecture and talk to those people about the subjects which applied to their locality. ``Acres of Diamonds''--the idea--has continuously been precisely the same. The idea is that in this country of ours every man has the opportunity to make more of himself than he does in his own environment, with his own skill, with his own energy, and with his own friends.

RUSSELL H. CONWELL.

ACRES OF DIAMONDS

WHEN going down the Tigris and Euphrates rivers many years ago with a party of English travelers I found myself under the direction of an old Arab guide whom we hired up at Baghdad, and I have often thought how that guide resembled our barbers in certain mental characteristics. He thought that it was not only his duty to guide us down those rivers, and do what he was paid for doing, but also to entertain us with stories curious and weird, ancient and modern, strange and familiar. Many of them I have forgotten, and I am glad I have, but there is one I shall never forget.

The old guide was leading my camel by its halter along the banks of those ancient rivers, and he told me story after story until I grew weary of his story-telling and ceased to listen. I have never been irritated with that guide when he lost his temper as I ceased listening. But I remember that he took off his Turkish cap and swung it in a circle to get my attention. I could see it through the corner of my eye, but I determined not to look straight at him for fear he would tell another story. But although I am not a woman, I did finally look, and as soon as I did he went right into another story.

Said he, ``I will tell you a story now which I reserve for my particular friends." When he emphasized the words ``particular friends," I listened, and I have ever been glad I did. I really feel devoutly thankful, that there are 1,674 young men who have been carried through college by this lecture who are also glad that I did listen. The old guide told me that there once lived not far from the River Indus an ancient Persian by the name of Ali Hafed. He said that Ali Hafed owned a very large farm, that he had orchards, grain-fields, and gardens; that he had money at interest, and was a wealthy and contented man. He was contented because he was wealthy, and wealthy because he was contented. One day there visited that old Persian farmer one of these ancient Buddhist priests, one of the wise men of the East. He sat down by the fire and told the old farmer how this world of ours was made. He said that this world was once a mere bank of fog, and that the Almighty thrust His finger into this bank of fog, and began slowly to move His finger around, increasing the speed until at last He whirled this bank of fog into a solid ball of fire. Then it went rolling through the universe, burning its way through other banks of fog, and condensed the moisture without, until it fell in floods of rain upon its hot surface, and cooled the outward crust. Then the internal fires bursting outward through the crust threw up the mountains and hills, the valleys, the plains and prairies of this wonderful world of ours. If this internal molten mass came bursting out and cooled very quickly it became granite; less quickly copper, less quickly silver, less quickly gold, and, after gold, diamonds were made.

Said the old priest, ``A diamond is a congealed drop of sunlight." Now that is literally scientifically true, that a diamond is an actual deposit of carbon from the sun. The old priest told Ali Hafed that if he had one diamond the size of his thumb he could purchase the county, and if he had a mine of diamonds he could place his children upon thrones through the influence of their great wealth.

Ali Hafed heard all about diamonds, how much they were worth, and went to his bed that night a poor man. He had not lost anything, but he was poor because he was discontented, and discontented because he feared he was poor. He said, ``I want a mine of diamonds," and he lay awake all night.

Early in the morning he sought out the priest. I know by experience that a priest is very cross when awakened early in the morning, and when he shook that old priest out of his dreams, Ali Hafed said to him:

``Will you tell me where I can find diamonds?"

``Diamonds! What do you want with diamonds?" ``Why, I wish to be immensely rich." ``Well, then, go along and find them. That is all you have to do; go and find them, and then you have them." ``But I don't know where to go." ``Well, if you will find a river that runs through white sands, between high mountains, in those white sands you will always find diamonds." ``I don't believe there is any such river." ``Oh yes, there are plenty of them. All you have to do is to go and find them, and then you have them." Said Ali Hafed, ``I will go."

So he sold his farm, collected his money, left his family in charge of a neighbor, and away he went in search of diamonds. He began his search, very properly to my mind, at the Mountains of the Moon. Afterward he came around into Palestine, then wandered on into Europe, and at last when his money was all spent and he was in rags, wretchedness, and poverty, he stood on the shore of that bay at Barcelona, in Spain, when a great tidal wave came rolling in between the pillars of Hercules, and the poor, afflicted, suffering, dying man could not resist the awful temptation to cast himself into that incoming tide, and he sank beneath its foaming crest, never to rise in this life again.

When that old guide had told me that awfully sad story he stopped the camel I was riding on and went back to fix the baggage that was coming off another camel, and I had an opportunity to muse over his story while he was gone. I remember saying to myself, ``Why did he reserve that story for his `particular friends'?" There seemed to be no beginning, no middle, no end, nothing to it. That was the first story I had ever heard told in my life, and would be the first one I ever read, in which the hero was killed in the first chapter. I had but one chapter of that story, and the hero was dead.

When the guide came back and took up the halter of my camel, he went right ahead with the story, into the second chapter, just as though there had been no break. The man who purchased Ali Hafed's farm one day led his camel into the garden to drink, and as that camel put its nose into the shallow water of that garden brook, Ali Hafed's successor noticed a curious flash of light from the white sands of the stream. He pulled out a black stone having an eye of light reflecting all the hues of the rainbow. He took the pebble into the house and put it on the mantel which covers the central fires, and forgot all about it.

A few days later this same old priest came in to visit Ali Hafed's successor, and the moment he opened that drawing-room door he saw that flash of light on the mantel, and he rushed up to it, and shouted: ``Here is a diamond! Has Ali Hafed returned?'' ``Oh no, Ali Hafed has not returned, and that is not a diamond. That is nothing but a stone we found right out here in our own garden.'' ``But,'' said the priest, ``I tell you I know a diamond when I see it. I know positively that is a diamond.''

Then together they rushed out into that old garden and stirred up the white sands with their fingers, and lo! there came up other

38

more beautiful and valuable gems than the first. ``Thus,'' said the guide to me, and, friends, it is historically true, ``was discovered the diamond-mine of Golconda, the most magnificent diamond-mine in all the history of mankind, excelling the Kimberly itself. The Kohinoor, and the Orloff of the crown jewels of England and Russia, the largest on earth, came from that mine.''

When that old Arab guide told me the second chapter of his story, he then took off his Turkish cap and swung it around in the air again to get my attention to the moral. Those Arab guides have morals to their stories, although they are not always moral. As he swung his hat, he said to me, ``Had Ali Hafed remained at home and dug in his own cellar, or underneath his own wheat-fields, or in his own garden, instead of wretchedness, starvation, and death by suicide in a strange land, he would have had `acres of diamonds.' For every acre of that old farm, yes, every shovelful, afterward revealed gems which since have decorated the crowns of monarchs.''

When he had added the moral to his story I saw why he reserved it for ``his particular friends.'' But I did not tell him I could see it. It was that mean old Arab's way of going around a thing like a lawyer, to say indirectly what he did not dare say directly, that ``in his private opinion there was a certain young man then traveling down the Tigris River that might better be at home in America.'' I did not tell him I could see that, but I told him his story reminded me of one, and I told it to him quick, and I think I will tell it to you.

I told him of a man out in California in 1847 who owned a ranch. He heard they had discovered gold in southern California, and so with a passion for gold he sold his ranch to Colonel Sutter, and away he went, never to come back. Colonel Sutter put a mill upon a stream that ran through that ranch, and one day his little girl brought some wet sand from the raceway into their home and sifted it through her fingers before the fire, and in that falling sand a visitor saw the first shining scales of real gold that were ever discovered in California. The man who had owned that ranch wanted gold, and he could have secured it for the mere taking. Indeed, thirty-eight millions of dollars has been taken out of a very few acres since then. About eight years ago I delivered this lecture in a city that stands on that farm, and they told me that a one-third owner for years and years had been getting one hundred and twenty dollars in gold every fifteen minutes, sleeping or waking, without taxation. You and I would enjoy an income like that--if we didn't have to pay an income tax.

But a better illustration really than that occurred here in our own Pennsylvania. If there is anything I enjoy above another on the platform, it is to get one of these German audiences in Pennsylvania before me, and fire that at them, and I enjoy it to-night. There was a man living in Pennsylvania, not unlike some

40

Pennsylvanians you have seen, who owned a farm, and he did with that farm just what I should do with a farm if I owned one in Pennsylvania--he sold it. But before he sold it he decided to secure employment collecting coal-oil for his cousin, who was in the business in Canada, where they first discovered oil on this continent. They dipped it from the running streams at that early time. So this Pennsylvania farmer wrote to his cousin asking for employment. You see, friends, this farmer was not altogether a foolish man. No, he was not. He did not leave his farm until he had something else to do. _*Of all the simpletons the stars shine on I don't know of a worse one than the man who leaves one job before he has gotten another_. That has especial reference to my profession, and has no reference whatever to a man seeking a divorce. When he wrote to his cousin for employment, his cousin replied, ``I cannot engage you because you know nothing about the oil business."

Well, then the old farmer said, ``I will know," and with most commendable zeal (characteristic of the students of Temple University) he set himself at the study of the whole subject. He began away back at the second day of God's creation when this world was covered thick and deep with that rich vegetation which since has turned to the primitive beds of coal. He studied the subject until he found that the drainings really of those rich beds of coal furnished the coal-oil that was worth pumping, and then he found how it came up with the living springs. He studied until he knew what it looked like, smelled like, tasted like, and how to refine it. Now said he in his letter to his cousin, ``I understand the oil business." His cousin answered, ``All right, come on."

So he sold his farm, according to the county record, for $833 (even money, ``no cents"). He had scarcely gone from that place before the man who purchased the spot went out to arrange for the watering of the cattle. He found the previous owner had gone out years before and put a plank across the brook back of the barn, edgewise into the surface of the water just a few inches. The purpose of that plank at that sharp angle across the brook was to throw over to the other bank a dreadful-looking scum through which the cattle would not put their noses. But with that plank there to throw it all over to one side, the cattle would drink below, and thus that man who had gone to Canada had been himself damming back for twenty-three years a flood of coal-oil which the state geologists of Pennsylvania declared to us ten years later was even then worth a hundred millions of dollars to our state, and four years ago our geologist declared the discovery to be worth to our state a thousand millions of dollars. The man who owned that territory on which the city of Titusville now stands, and those Pleasantville valleys, had studied the subject from the second day of God's creation clear down to the present time. He studied it until he knew all about it, and yet he is said to have sold the whole of it for $833, and again I say, ``no sense."

But I need another illustration. I found it in Massachusetts, and I am sorry I did because that is the state I came from. This young man in Massachusetts furnishes just another phase of my thought. He went to Yale College and studied mines and mining, and became such an adept as a mining engineer that he was employed by the authorities of the university to train students who were behind their classes. During his senior year he earned $15 a week for doing that work. When he graduated they raised his pay from $15 to $45 a week, and offered him a professorship, and as soon as they did he went right home to his mother.

*If they had raised that boy's pay from $15 to $15.60 he would have stayed and been proud of the place, but when they put it up to $45 at one leap, he said, ``Mother, I won't work for $45 a week. The idea of a man with a brain like mine working for $45 a week! Let's go out in California and stake out gold-mines and silver-mines, and be immensely rich."

Said his mother, ``Now, Charlie, it is just as well to be happy as it is to be rich."

``Yes," said Charlie, ``but it is just as well to be rich and happy, too." And they were both right about it. As he was an only son and she a widow, of course he had his way. They always do.

43

They sold out in Massachusetts, and instead of going to California they went to Wisconsin, where he went into the employ of the Superior Copper Mining Company at $15 a week again, but with the proviso in his contract that he should have an interest in any mines he should discover for the company. I don't believe he ever discovered a mine, and if I am looking in the face of any stockholder of that copper company you wish he had discovered something or other. I have friends who are not here because they could not afford a ticket, who did have stock in that company at the time this young man was employed there. This young man went out there, and I have not heard a word from him. I don't know what became of him, and I don't know whether he found any mines or not, but I don't believe he ever did.

But I do know the other end of the line. He had scarcely gotten out of the old homestead before the succeeding owner went out to dig potatoes. The potatoes were already growing in the ground when he bought the farm, and as the old farmer was bringing in a basket of potatoes it hugged very tight between the ends of the stone fence. You know in Massachusetts our farms are nearly all stone wall. There you are obliged to be very economical of front gateways in order to have some place to put the stone. When that basket hugged so tight he set it down on the ground, and then dragged on one side, and pulled on the other side, and as he was dragging that basket through this farmer noticed in the upper and outer corner of that stone wall, right next the gate, a block of native silver eight inches square. That professor of mines, mining, and mineralogy who knew so much about the subject that he would not work for $45 a week, when he sold that homestead in Massachusetts sat right on that silver to make the bargain. He was born on that homestead, was brought up there, and had gone back and forth rubbing the stone with his sleeve until it reflected his countenance, and seemed to say, ``Here is a hundred thousand dollars right down here just for the taking." But he would not take it. It was in a home in Newburyport, Massachusetts, and there was no silver there, all away off--well, I don't know where, and he did not, but somewhere else, and he was a professor of mineralogy.

My friends, that mistake is very universally made, and why should we even smile at him. I often wonder what has become of him. I do not know at all, but I will tell you what I ``guess" as a Yankee. I guess that he sits out there by his fireside to-night with his friends gathered around him, and he is saying to them something like this: ``Do you know that man Conwell who lives in Philadelphia?" ``Oh yes, I have heard of him." ``Do you know that man Jones that lives in Philadelphia?" ``Yes, I have heard of him, too."

Then he begins to laugh, and shakes his sides and says to his friends, ``Well, they have done just the same thing I did, precisely''--and that spoils the whole joke, for you and I have done the same thing he did, and while we sit here and laugh at him he has a better right to sit out there and laugh at us. I know I have made the same mistakes, but, of course, that does not make any difference, because we don't expect the same man to preach and practice, too.

As I come here to-night and look around this audience I am seeing again what through these fifty years I have continually seen-men that are making precisely that same mistake. I often wish I could see the younger people, and would that the Academy had been filled to-night with our high- school scholars and our grammar-school scholars, that I could have them to talk to. While I would have preferred such an audience as that, because they are most susceptible, as they have not grown up into their prejudices as we have, they have not gotten into any custom that they cannot break, they have not met with any failures as we have; and while I could perhaps do such an audience as that more good than I can do grown- up people, yet I will do the best I can with the material I have. I say to you that you have ``acres of diamonds'' in Philadelphia right where you now live. ``Oh,'' but you will say, ``you cannot know much about your city if you think there are any `acres of diamonds' here.''

I was greatly interested in that account in the newspaper of the young man who found that diamond in North Carolina. It was one of the purest diamonds that has ever been discovered, and it has several predecessors near the same locality. I went to a distinguished professor in mineralogy and asked him where he thought those diamonds came from. The professor secured the map of the geologic formations of our continent, and traced it. He said it went either through the underlying carboniferous strata adapted for such production, westward through Ohio and the Mississippi, or in more probability came eastward through Virginia and up the shore of the Atlantic Ocean. It is a fact that the diamonds were there, for they have been discovered and sold; and that they were carried down there during the drift period, from some northern locality. Now who can say but some person going down with his drill in Philadelphia will find some trace of a diamond-mine yet down here? Oh, friends! you cannot say that you are not over one of the greatest diamond-mines in the world, for such a diamond as that only comes from the most profitable mines that are found on earth.

But it serves simply to illustrate my thought, which I emphasize by saying if you do not have the actual diamond-mines literally you have all that they would be good for to you. Because now that the Queen of England has given the greatest compliment ever conferred upon American woman for her attire because she did not appear with any jewels at all at the late reception in England, it has almost done away with the use of diamonds anyhow. All you would care for would be the few you would wear if you wish to be modest, and the rest you would sell for money.

Now then, I say again that the opportunity to get rich, to attain unto great wealth, is here in Philadelphia now, within the reach of almost every man and woman who hears me speak to- night, and I mean just what I say. I have not come to this platform even under these circumstances to recite something to you. I have come to tell you what in God's sight I believe to be the truth, and if the years of life have been of any value to me in the attainment of common sense, I know I am right; that the men and women sitting here, who found it difficult perhaps to buy a ticket to this lecture or gathering to-night, have within their reach ``acres of diamonds,'' opportunities to get largely wealthy. There never was a place on earth more adapted than the city of Philadelphia to-day, and never in the history of the world did a poor man without capital have such an opportunity to get rich quickly and honestly as he has now in our city. I say it is the truth, and I want you to accept it as such; for if you think I have come to simply recite something, then I would better not be here. I have no time to waste in any such talk, but to say the things I believe, and unless some of you get richer for what I am saying to-night my time is wasted.

I say that you ought to get rich, and it is your duty to get rich.

How many of my pious brethren say to me, ``Do you, a Christian minister, spend your time going up and down the country advising young people to get rich, to get money?'' ``Yes, of course I do.'' They say, ``Isn't that awful! Why don't you preach the gospel instead of preaching about man's making money?'' ``Because to make money honestly is to preach the gospel.'' That is the reason. The men who get rich may be the most honest men you find in the community.

``Oh,'' but says some young man here to-night, ``I have been told all my life that if a person has money he is very dishonest and dishonorable and mean and contemptible. ``My friend, that is the reason why you have none, because you have that idea of people. The foundation of your faith is altogether false. Let me say here clearly, and say it briefly, though subject to discussion which I have not time for here, ninety-eight out of one hundred of the rich men of America are honest. That is why they are rich. That is why they are trusted with money. That is why they carry on great enterprises and find plenty of people to work with them. It is because they are honest men.

Says another young man, ``I hear sometimes of men that get millions of dollars dishonestly.'' Yes, of course you do, and so do I. But they are so rare a thing in fact that the newspapers talk about them all the time as a matter of news until you get the idea that all the other rich men got rich dishonestly.

My friend, you take and drive me--if you furnish the auto--out into the suburbs of Philadelphia, and introduce me to the people who own their homes around this great city, those beautiful homes with gardens and flowers, those magnificent homes so lovely in their art, and I will introduce you to the very best people in character as well as in enterprise in our city, and you know I will. A man is not really a true man until he owns his own home, and they that own their homes are made more honorable and honest and pure, and true and economical and careful, by owning the home.

For a man to have money, even in large sums, is not an inconsistent thing. We preach against covetousness, and you know we do, in the pulpit, and oftentimes preach against it so long and use the terms about ``filthy lucre'' so extremely that Christians get the idea that when we stand in the pulpit we believe it is wicked for any man to have money--until the collection-basket goes around, and then we almost swear at the people because they don't give more money. Oh, the inconsistency of such doctrines as that!

Money is power, and you ought to be reasonably ambitious to have it. You ought because you can do more good with it than you could without it. Money printed your Bible, money builds your churches, money sends your missionaries, and money pays your preachers, and you would not have many of them, either, if you did not pay them. I am always willing that my church should raise my salary, because the church that pays the largest salary always raises it the easiest. You never knew an exception to it in your life. The man who gets the largest salary can do the most good with the power that is furnished to him. Of course he can if his spirit be right to use it for what it is given to him.

I say, then, you ought to have money. If you can honestly attain unto riches in Philadelphia, it is your Christian and godly duty to do so. It is an awful mistake of these pious people to think you must be awfully poor in order to be pious.

Some men say, ``Don't you sympathize with the poor people?" Of course I do, or else I would not have been lecturing these years. I won't give in but what I sympathize with the poor, but the number of poor who are to be sympathized with is very small. To sympathize with a man whom God has punished for his sins, thus to help him when God would still continue a just punishment, is to do wrong, no doubt about it, and we do that more than we help those who are deserving. While we should sympathize with God's poor--that is, those who cannot help themselves-- let us remember there is not a poor person in the United States who was not made poor by his own shortcomings, or by the shortcomings of some one else. It is all wrong to be poor, anyhow. Let us give in to that argument and pass that to one side.

A gentleman gets up back there, and says, ``Don't you think there are some things in this world that are better than money?" Of course I do, but I am talking about money now. Of course there are some things higher than money. Oh yes, I know by the grave that has left me standing alone that there are some things in this world that are higher and sweeter and purer than money. Well do I know there are some things higher and grander than gold. Love is the grandest thing on God's earth, but fortunate the lover who has plenty of money. Money is power, money is force, money will do good as well as harm. In the hands of good men and women it could accomplish, and it has accomplished, good.

I hate to leave that behind me. I heard a man get up in a prayer-meeting in our city and thank the Lord he was ``one of God's poor." Well, I wonder what his wife thinks about that? She earns all the money that comes into that house, and he smokes a part of that on the veranda. I don't want to see any more of the Lord's poor of that kind, and I don't believe the Lord does. And yet there are some people who think in order to be pious you must be awfully poor and awfully dirty. That does not follow at all. While we sympathize with the poor, let us not teach a doctrine like that.

Yet the age is prejudiced against advising a Christian man (or, as a Jew would say, a godly man) from attaining unto wealth. The prejudice is so universal and the years are far enough back, I think, for me to safely mention that years ago up at Temple

University there was a young man in our theological school who thought he was the only pious student in that department. He came into my office one evening and sat down by my desk, and said to me: ``Mr. President, I think it is my duty sir, to come in and labor with you." ``What has happened now?" Said he, ``I heard you say at the Academy, at the Peirce School commencement, that you thought it was an honorable ambition for a young man to desire to have wealth, and that you thought it made him temperate, made him anxious to have a good name, and made him industrious. You spoke about man's ambition to have money helping to make him a good man. Sir, I have come to tell you the Holy Bible says that `money is the root of all evil.' "

I told him I had never seen it in the Bible, and advised him to go out into the chapel and get the Bible, and show me the place. So out he went for the Bible, and soon he stalked into my office with the Bible open, with all the bigoted pride of the narrow sectarian, or of one who founds his Christianity on some misinterpretation of Scripture. He flung the Bible down on my desk, and fairly squealed into my ear: ``There it is, Mr. President; you can read it for yourself." I said to him: ``Well, young man, you will learn when you get a little older that you cannot trust another denomination to read the Bible for you. You belong to another denomination. You are taught in the theological school, however, that emphasis is exegesis. Now, will you take that Bible and read it yourself, and give the proper emphasis to it?"

He took the Bible, and proudly read, `` `The love of money is the root of all evil.' "

Then he had it right, and when one does quote aright from that same old Book he quotes the absolute truth. I have lived through fifty years of the mightiest battle that old Book has ever fought, and I have lived to see its banners flying free; for never in the history of this world did the great minds of earth so universally agree that the Bible is true--all true--as they do at this very hour.

So I say that when he quoted right, of course he quoted the absolute truth. ``The love of money is the root of all evil." He who tries to attain unto it too quickly, or dishonestly, will fall into many snares, no doubt about that. The love of money. What is that? It is making an idol of money, and idolatry pure and simple everywhere is condemned by the Holy Scriptures and by man's common sense. The man that worships the dollar instead of thinking of the purposes for which it ought to be used, the man who idolizes simply money, the miser that hordes his money in the cellar, or hides it in his stocking, or refuses to invest it where it will do the world good, that man who hugs the dollar until the eagle squeals has in him the root of all evil.

I think I will leave that behind me now and answer the question of nearly all of you who are asking, ``Is there opportunity to get rich in Philadelphia?'' Well, now, how simple a thing it is to see where it is, and the instant you see where it is it is yours. Some old gentleman gets up back there and says, ``Mr. Conwell, have you lived in Philadelphia for thirty-one years and don't know that the time has gone by when you can make anything in this city?'' ``No, I don't think it is.'' ``Yes, it is; I have tried it.'' ``What business are you in?'' ``I kept a store here for twenty years, and never made over a thousand dollars in the whole twenty years.''

``Well, then, you can measure the good you have been to this city by what this city has paid you, because a man can judge very well what he is worth by what he receives; that is, in what he is to the world at this time. If you have not made over a thousand dollars in twenty years in Philadelphia, it would have been better for Philadelphia if they had kicked you out of the city nineteen years and nine months ago. A man has no right to keep a store in Philadelphia twenty years and not make at least five hundred thousand dollars even though it be a corner grocery up-town.' You say, ``You cannot make five thousand dollars in a store now.'' Oh, my friends, if you will just take only four blocks around you, and find out what the people want and what you ought to supply and set them down with your pencil and figure up the profits you would make if you did supply them, you would very soon see it. There is wealth right within the sound of your voice.

Some one says: ``You don't know anything about business. A preacher never knows a thing about business.'' Well, then, I will have to prove that I am an expert. I don't like to do this, but I have to do it because my testimony will not be taken if I am not an expert. My father kept a country store, and if there is any place under the stars where a man gets all sorts of experience in every kind of mercantile transactions, it is in the country store. I am not proud of my experience, but sometimes when my father was away he would leave me in charge of the store, though fortunately for him that was not very often. But this did occur many times, friends: A man would come in the store, and say to me, ``Do you keep jack knives?'' ``No, we don't keep jack-knives,'' and I went off whistling a tune. What did I care about that man, anyhow? Then another farmer would come in and say, ``Do you keep jack knives?'' ``No, we don't keep jack-knives.'' Then I went away and whistled another tune. Then a third man came right in the same door and said, ``Do you keep jack-knives?'' ``No. Why is every one around here asking for jack-knives? Do you suppose we are keeping this store to supply the whole neighborhood with jack-knives?'' Do you carry on your store like that in Philadelphia? The difficulty was I had not then learned that the foundation of godliness and the foundation principle of success in business are both the same precisely. The man who says, ``I cannot carry my religion into business'' advertises himself either as being an imbecile in business, or on the road to bankruptcy, or a thief, one

56

of the three, sure. He will fail within a very few years. He certainly will if he doesn't carry his religion into business. If I had been carrying on my father's store on a Christian plan, godly plan, I would have had a jack-knife for the third man when he called for it. Then I would have actually done him a kindness, and I would have received a reward myself, which it would have been my duty to take.

There are some over-pious Christian people who think if you take any profit on anything you sell that you are an unrighteous man. On the contrary, you would be a criminal to sell goods for less than they cost. You have no right to do that. You cannot trust a man with your money who cannot take care of his own. You cannot trust a man in your family that is not true to his own wife. You cannot trust a man in the world that does not begin with his own heart, his own character, and his own life. It would have been my duty to have furnished a jack-knife to the third man, or the second, and to have sold it to him and actually profited myself. I have no more right to sell goods without making a profit on them than I have to overcharge him dishonestly beyond what they are worth. But I should so sell each bill of goods that the person to whom I sell shall make as much as I make.

To live and let live is the principle of the gospel, and the principle of every-day common sense. Oh, young man, hear me; live as you go along. Do not wait until you have reached my years before you begin to enjoy anything of this life. If I had the millions back, or fifty cents of it, which I have tried to earn in these years, it would not do me anything like the good that it does me now in this almost sacred presence to- night. Oh, yes, I am paid over and over a hundredfold to-night for dividing as I have tried to do in some measure as I went along through the years. I ought not speak that way, it sounds egotistic, but I am old enough now to be excused for that. I should have helped my fellow-men, which I have tried to do, and every one should try to do, and get the happiness of it. The man who goes home with the sense that he has stolen a dollar that day, that he has robbed a man of what was his honest due, is not going to sweet rest. He arises tired in the morning, and goes with an unclean conscience to his work the next day. He is not a successful man at all, although he may have laid up millions. But the man who has gone through life dividing always with his fellow-men, making and demanding his own rights and his own profits, and giving to every other man his rights and profits, lives every day, and not only that, but it is the royal road to great wealth. The history of the thousands of millionaires shows that to be the case.

The man over there who said he could not make anything in a store in Philadelphia has been carrying on his store on the wrong principle. Suppose I go into your store to-morrow morning and ask, "Do you know neighbor A, who lives one square away, at house No. 1240?" ``Oh yes, I have met him. He deals here at the corner store." ``Where did he come from?" ``I don't know." ``How many does he have in his family?" ``I don't know." ``What ticket does he vote?" ``I don't know." ``What church does he go to?" ``I don't know, and don't care. What are you asking all these questions for?"

If you had a store in Philadelphia would you answer me like that? If so, then you are conducting your business just as I carried on my father's business in Worthington, Massachusetts. You don't know where your neighbor came from when he moved to Philadelphia, and you don't care. If you had cared you would be a rich man now. If you had cared enough about him to take an interest in his affairs, to find out what he needed, you would have been rich. But you go through the world saying, ``No opportunity to get rich," and there is the fault right at your own door.

But another young man gets up over there and says, ``I cannot take up the mercantile business." (While I am talking of trade it applies to every occupation.) ``Why can't you go into the mercantile business?" ``Because I haven't any capital." Oh, the weak and dudish creature that can't see over its collar! It makes a person weak to see these little dudes standing around the corners and saying, ``Oh, if I had plenty of capital, how rich I would get." ``Young man, do you think you are going to get rich on capital?" ``Certainly." Well, I say, ``Certainly not." If your mother has plenty of money, and she will set you up in business, you will ``set her up in business,'' supplying you with capital.

The moment a young man or woman gets more money than he or she has grown to by practical experience, that moment he has gotten a curse. It is no help to a young man or woman to inherit money. It is no help to your children to leave them money, but if you leave them education, if you leave them Christian and noble character, if you leave them a wide circle of friends, if you leave them an honorable name, it is far better than that they should have money. It would be worse for them, worse for the nation, that they should have any money at all. Oh, young man, if you have inherited money, don't regard it as a help. It will curse you through your years, and deprive you of the very best things of human life. There is no class of people to be pitied so much as the inexperienced sons and daughters of the rich of our generation. I pity the rich man's son. He can never know the best things in life.

One of the best things in our life is when a young man has earned his own living, and when he becomes engaged to some lovely young woman, and makes up his mind to have a home of his own. Then with that same love comes also that divine inspiration toward better things, and he begins to save his money. He begins to leave off his bad habits and put money in the bank. When he has a few hundred dollars he goes out in the suburbs to look for a home. He goes to the savings-bank, perhaps, for half of the value, and then goes for his wife, and when he takes his bride over the threshold of that door for the first time he says in words of eloquence my voice can never touch: ``I have earned this home myself. It is all mine, and I divide with thee.'' That is the grandest moment a human heart may ever know.

But a rich man's son can never know that. He takes his bride into a finer mansion, it may be, but he is obliged to go all the way through it and say to his wife, ``My mother gave me that, my mother gave me that, and my mother gave me this,'' until his wife wishes she had married his mother. I pity the rich man's son.

The statistics of Massachusetts showed that not one rich man's son out of seventeen ever dies rich. I pity the rich man's sons unless they have the good sense of the elder Vanderbilt, which sometimes happens. He went to his father and said, ``Did you earn all your money?'' ``I did, my son. I began to work on a ferry-boat for twenty-five cents a day.'' ``Then,'' said his son, ``I will have none of your money,'' and he, too, tried to get employment on a ferry-boat that Saturday night. He could not get one there, but he did get a place for three dollars a week. Of course, if a rich man's son will do that, he will get the discipline of a poor boy that is worth more than a university education to any man. He would then be able to take care of the millions of his father. But as a rule the rich men will not let their sons do the very thing that made them great. As a rule, the rich man will not allow his son to work-

-and his mother! Why, she would think it was a social disgrace if her poor, weak, little lily-fingered, sissy sort of a boy had to earn his living with honest toil. I have no pity for such rich men's sons.

I remember one at Niagara Falls. I think I remember one a great deal nearer. I think there are gentlemen present who were at a great banquet, and I beg pardon of his friends. At a banquet here in Philadelphia there sat beside me a kind-hearted young man, and he said, ``Mr. Conwell, you have been sick for two or three years. When you go out, take my limousine, and it will take you up to your house on Broad Street." I thanked him very much, and perhaps I ought not to mention the incident in this way, but I follow the facts. I got on to the seat with the driver of that limousine, outside, and when we were going up I asked the driver, ``How much did this limousine cost?" ``Six thousand eight hundred, and he had to pay the duty on it." ``Well," I said, ``does the owner of this machine ever drive it himself?" At that the chauffeur laughed so heartily that he lost control of his machine. He was so surprised at the question that he ran up on the sidewalk, and around a corner lamp-post out into the street again. And when he got out into the street he laughed till the whole machine trembled. He said: ``He drive this machine! Oh, he would be lucky if he knew enough to get out when we get there."

I must tell you about a rich man's son at Niagara Falls. I came in from the lecture to the hotel, and as I approached the desk of the clerk there stood a millionaire's son from New York. He was an indescribable specimen of anthropologic potency. He had a skull-cap on one side of his head, with a gold tassel in the top of it, and a gold-headed cane under his arm with more in it than in his head. It is a very difficult thing to describe that young man. He wore an eye- glass that he could not see through, patent- leather boots that he could not walk in, and pants that he could not sit down in--dressed like a grasshopper. This human cricket came up to the clerk's desk just as I entered, adjusted his unseeing eye-glass, and spoke in this wise to the clerk. You see, he thought it was ``Hinglish, you know,'' to lisp. ``Thir, will you have the kindness to supply me with thome papah and enwelophs!'' The hotel clerk measured that man quick, and he pulled the envelopes and paper out of a drawer, threw them across the counter toward the young man, and then turned away to his books. You should have seen that young man when those envelopes came across that counter. He swelled up like a gobbler turkey, adjusted his unseeing eye- glass, and yelled: ``Come right back here. Now thir, will you order a thervant to take that papah and enwelophs to yondah dethk.'' Oh, the poor, miserable, contemptible American monkey! He could not carry paper and envelopes twenty feet. I suppose he could not get his arms down to do it. I have no pity for such travesties upon human nature. If you have not capital, young man, I am glad of it. What you need is common sense, not copper cents.

The best thing I can do is to illustrate by actual facts well-known to you all. A. T. Stewart, a poor boy in New York, had $1.50 to begin life on. He lost 87 <1/2> cents of that on the very first venture. How fortunate that young man who loses the first time he gambles. That boy said, ``I will never gamble again in business,'' and he never did. How came he to lose 87 <1/2> cents? You probably all know the story how he lost it--because he bought some needles, threads, and buttons to sell which people did not want, and had them left on his hands, a dead loss. Said the boy, ``I will not lose any more money in that way.'' Then he went around first to the doors and asked the people what they did want. Then when he had found out what they wanted he invested his 62 <1/2> cents to supply a known demand. Study it wherever you choose--in business, in your profession, in your housekeeping, whatever your life, that one thing is the secret of success. You must first know the demand. You must first know what people need, and then invest yourself where you are most needed. A. T. Stewart went on that principle until he was worth what amounted afterward to forty millions of dollars, owning the very store in which Mr. Wanamaker carries on his great work in New York. His fortune was made by his losing something, which taught him the great lesson that he must only invest himself or his money in something that people need. When will you salesmen learn it? When will you manufacturers learn that you must know the changing needs of humanity if you would succeed in life?

Apply yourselves, all you Christian people, as manufacturers or merchants or workmen to supply that human need. It is a great principle as broad as humanity and as deep as the Scripture itself.

The best illustration I ever heard was of John Jacob Astor. You know that he made the money of the Astor family when he lived in New York. He came across the sea in debt for his fare. But that poor boy with nothing in his pocket made the fortune of the Astor family on one principle. Some young man here to-night will say, ``Well they could make those fortunes over in New York but they could not do it in Philadelphia!'' My friends, did you ever read that wonderful book of Riss (his memory is sweet to us because of his recent death), wherein is given his statistical account of the records taken in 1889 of 107 millionaires of New York. If you read the account you will see that out of the 107 millionaires only seven made their money in New York. Out of the 107 millionaires worth ten million dollars in real estate then, 67 of them made their money in towns of less than 3,500 inhabitants. The richest man in this country to-day, if you read the real-estate values, has never moved away from a town of 3,500 inhabitants. It makes not so much difference where you are as who you are. But if you cannot get rich in Philadelphia you certainly cannot do it in New York.

Now John Jacob Astor illustrated what can be done anywhere. He had a mortgage once on a millinery-store, and they could not sell bonnets enough to pay the interest on his money. So he foreclosed that mortgage, took possession of the store, and went into partnership with the very same people, in the same store, with the same capital. He did not give them a dollar of capital. They had to sell goods to get any money. Then he left them alone in the store just as they had been before, and he went out and sat down on a bench in the park in the shade. What was John Jacob Astor doing out there, and in partnership with people who had failed on his own hands? He had the most important and, to my mind, the most pleasant part of that partnership on his hands. For as John Jacob Astor sat on that bench he was watching the ladies as they went by; and where is the man who would not get rich at that business? As he sat on the bench if a lady passed him with her shoulders back and head up, and looked straight to the front, as if she did not care if all the world did gaze on her, then he studied her bonnet, and by the time it was out of sight he knew the shape of the frame, the color of the trimmings, and the crinklings in the feather. I sometimes try to describe a bonnet, but not always. I would not try to describe a modern bonnet. Where is the man that could describe one? This aggregation of all sorts of driftwood stuck on the back of the head, or the side of the neck, like a rooster with only one tail feather left. But in John Jacob Astor's day there was some art about the millinery business, and he went to the millinery-store and said to them: ``Now put into the show-window just such a bonnet as I describe to you, because I have already seen a lady who likes such a bonnet. Don't make up any more until I come back." Then he went out and sat down again, and another lady passed him of a different form, of different complexion, with a different shape and color of bonnet. ``Now," said he, ``put such a bonnet as that in the show window." He did not fill his show-window up town with a lot of hats and bonnets to drive people away, and then sit on the back stairs and bawl because people went to Wanamaker's to trade. He did not have a hat or a bonnet in that show-window but what some lady liked before it was made up. The tide of custom began immediately to turn in, and that has been the foundation of the greatest store in New York in that line, and still exists as one of three stores. Its fortune was made by John Jacob Astor after they had failed in business, not by giving them any more money, but by finding out what the ladies liked for bonnets before they wasted any material in making them up. I tell you if a man could foresee the millinery business he could foresee anything under heaven!

Suppose I were to go through this audience to-night and ask you in this great manufacturing city if there are not opportunities to get rich in manufacturing. ``Oh yes," some young man says, ``there are opportunities here still if you build with some trust and if you have two or three millions of dollars to begin with as capital." Young man, the history of the breaking up of the trusts

67

by that attack upon ``big business'' is only illustrating what is now the opportunity of the smaller man. The time never came in the history of the world when you could get rich so quickly manufacturing without capital as you can now.

But you will say, ``You cannot do anything of the kind. You cannot start without capital.'' Young man, let me illustrate for a moment. I must do it. It is my duty to every young man and woman, because we are all going into business very soon on the same plan. Young man, remember if you know what people need you have gotten more knowledge of a fortune than any amount of capital can give you.

There was a poor man out of work living in Hingham, Massachusetts. He lounged around the house until one day his wife told him to get out and work, and, as he lived in Massachusetts, he obeyed his wife. He went out and sat down on the shore of the bay, and whittled a soaked shingle into a wooden chain. His children that evening quarreled over it, and he whittled a second one to keep peace. While he was whittling the second one a neighbor came in and said: ``Why don't you whittle toys and sell them? You could make money at that.'' ``Oh,'' he said, ``I would not know what to make.'' ``Why don't you ask your own children right here in your own house what to make?'' ``What is the use of trying that?'' said the carpenter. ``My children are different from other people's children.'' (I used to see people like that when I taught school.) But he acted upon the hint, and the next morning when Mary came down the stairway, he asked, ``What do you want for a toy?'' She began to tell him she would like a doll's bed, a doll's washstand, a doll's carriage, a little doll's umbrella, and went on with a list of things that would take him a lifetime to supply. So, consulting his own children, in his own house, he took the firewood, for he had no money to buy lumber, and whittled those strong, unpainted Hingham toys that were for so many years known all over the world. That man began to make those toys for his own children, and then made copies and sold them through the boot-and-shoe store next door. He began to make a little money, and then a little more, and Mr. Lawson, in his *Frenzied Finance* says that man is the richest man in old Massachusetts, and I think it is the truth. And that man is worth a hundred millions of dollars to-day, and has been only thirty-four years making it on that one principle--that one must judge that what his own children like at home other people's children would like in their homes, too; to judge the human heart by oneself, by one's wife or by one's children. It is the royal road to success in manufacturing. ``Oh,'' but you say, ``didn't he have any capital?'' Yes, a penknife, but I don't know that he had paid for that.

I spoke thus to an audience in New Britain, Connecticut, and a lady four seats back went home and tried to take off her collar, and the collar- button stuck in the buttonhole. She threw it out and said, ``I am going to get up something better than that to put on collars.'' Her husband said: ``After what Conwell said to-night, you see there is a need of an improved collar-fastener that is easier to handle. There is a human need; there is a great fortune. Now, then, get up a collar-button and get rich.'' He made fun of her, and consequently made fun of me, and that is one of the saddest things which comes over me like a deep cloud of midnight sometimes--although I have worked so hard for more than half a century, yet how little I have ever really done. Notwithstanding the greatness and the handsomeness of your compliment to-night, I do not believe there is one in ten of you that is going to make a million of dollars because you are here to-night; but it is not my fault, it is yours. I say that sincerely. What is the use of my talking if people never do what I advise them to do? When her husband ridiculed her, she made up her mind she would make a better collar-button, and when a woman makes up her mind ``she will,'' and does not say anything about it, she does it. It was that New England woman who invented the snap button which you can find anywhere now. It was first a collar-button with a spring cap attached to the outer side. Any of you who wear modern waterproofs know the button that simply pushes together, and when you unbutton it you simply pull it apart. That is the button to which I refer, and which she invented. She afterward invented several other buttons, and then invested in more, and then was taken into partnership with great factories. Now that woman goes over the sea every summer in her private steamship--yes, and takes her husband with her! If her husband were to die, she would have money enough left now to buy a foreign duke or count or some such title as that at the latest quotations.

Now what is my lesson in that incident? It is this: I told her then, though I did not know her, what I now say to you, ``Your wealth is too near to you. You are looking right over it''; and she had to look over it because it was right under her chin.

I have read in the newspaper that a woman never invented anything. Well, that newspaper ought to begin again. Of course, I do not refer to gossip--I refer to machines--and if I did I might better include the men. That newspaper could never appear if women had not invented something. Friends, think. Ye women, think! You say you cannot make a fortune because you are in some laundry, or running a sewing-machine, it may be, or walking before some loom, and yet you can be a millionaire if you will but follow this almost infallible direction.

When you say a woman doesn't invent anything, I ask, Who invented the Jacquard loom that wove every stitch you wear? Mrs. Jacquard. The printer's roller, the printing-press, were invented by farmers' wives. Who invented the cotton-gin of the South that enriched our country so amazingly? Mrs. General Greene invented the cotton- gin and showed the idea to Mr. Whitney, and he, like a man, seized it. Who was it that invented the sewing-machine? If I would go to school to- morrow and ask your children they would say, ``Elias Howe."

He was in the Civil War with me, and often in my tent, and I often heard him say that he worked fourteen years to get up that sewing-machine. But his wife made up her mind one day that they would starve to death if there wasn't something or other invented pretty soon, and so in two hours she invented the sewing-machine. Of course he took out the patent in his name. Men always do that. Who was it that invented the mower and the reaper? According to Mr. McCormick's confidential communication, so recently published, it was a West Virginia woman, who, after his father and he had failed altogether in making a reaper and gave it up, took a lot of shears and nailed them together on the edge of a board, with one shaft of each pair loose, and then wired them so that when she pulled the wire one way it closed them, and when she pulled the wire the other way it opened them, and there she had the principle of the mowing-machine. If you look at a mowing-machine, you will see it is nothing but a lot of shears. If a woman can invent a mowing-machine, if a woman can invent a Jacquard loom, if a woman can invent a cotton-gin, if a woman can invent a trolley switch--as she did and made the trolleys possible; if a woman can invent, as Mr. Carnegie said, the great iron squeezers that laid the foundation of all the steel millions of the United States, ``we men'' can invent anything under the stars! I say that for the encouragement of the men.

Who are the great inventors of the world? Again this lesson comes before us. The great inventor sits next to you, or you are the person yourself. ``Oh,'' but you will say, ``I have never invented anything in my life.'' Neither did the great inventors until they discovered one great secret. Do you think it is a man with a head like a bushel measure or a man like a stroke of lightning? It is neither. The really great man is a plain, straightforward, every-day, common-sense man. You would not dream that he was a great inventor if you did not see something he had actually done. His neighbors do not regard him so great. You never see anything great over your back fence. You say there is no greatness among your neighbors. It is all away off somewhere else. Their greatness is ever so simple, so plain, so earnest, so practical, that the neighbors and friends never recognize it.

True greatness is often unrecognized. That is sure. You do not know anything about the greatest men and women. I went out to write the life of General Garfield, and a neighbor, knowing I was in a hurry, and as there was a great crowd around the front door, took me around to General Garfield's back door and shouted, ``Jim! Jim!'' And very soon ``Jim'' came to the door and let me in, and I wrote the biography of one of the grandest men of the nation, and yet he was just the same old ``Jim'' to his neighbor. If you know a great man in Philadelphia and you should meet him to-morrow, you would say, ``How are you, Sam?'' or ``Good morning, Jim.'' Of course you would. That is just what you would do.

One of my soldiers in the Civil War had been sentenced to death, and I went up to the White House in Washington--sent there for the first time in my life to see the President. I went into the waiting-room and sat down with a lot of others on the benches, and the secretary asked one after another to tell him what they wanted. After the secretary had been through the line, he went in, and then came back to the door and motioned for me. I went up to that anteroom, and the secretary said: ``That is the President's door right over there. Just rap on it and go right in.'' I never was so taken aback, friends, in all my life, never. The secretary himself made it worse for me, because he had told me how to go in and then went out another door to the left and shut that. There I was, in the hallway by myself before the President of the United States of America's door. I had been on fields of battle, where the shells did sometimes shriek and the bullets did sometimes hit me, but I always wanted to run. I have no sympathy with the old man who says, ``I would just as soon march up to the cannon's mouth as eat my dinner.'' I have no faith in a man who doesn't know enough to be afraid when he is being shot at. I never was so afraid when the shells came around us at Antietam as I was when I went into that room that day; but I finally mustered the courage-- I don't know how I ever did--and at arm's- length tapped on the door. The man inside did not help me at all, but yelled out, ``Come in and sit down!''

Well, I went in and sat down on the edge of a chair, and wished I were in Europe, and the man at the table did not look up. He was one of the world's greatest men, and was made great by one single rule. Oh, that all the young people of Philadelphia were before me now and I could say just this one thing, and that they would remember it. I would give a lifetime for the effect it would have on our city and on civilization. Abraham Lincoln's principle for greatness can be adopted by nearly all. This was his rule: Whatsoever he had to do at all, he put his whole mind into it and held it all there until that was all done. That makes men great almost anywhere. He stuck to those papers at that table and did not look up at me, and I sat there trembling. Finally, when he had put the string around his papers, he pushed them over to one side and looked over to me, and a smile came over his worn face. He said: ``I am a very busy man and have only a few minutes to spare. Now tell me in the fewest words what it is you want.'' I began to tell him, and mentioned the case, and he said: ``I have heard all about it and you do not need to say any more. Mr. Stanton was talking to me only a few days ago about that. You can go to the hotel and rest assured that the President never did sign an order to shoot a boy under twenty years of age, and never will. You can say that to his mother anyhow.''

Then he said to me, ``How is it going in the field?'' I said, ``We sometimes get discouraged.'' And he said: ``It is all right. We are going to win out now. We are getting very near the light. No man ought to wish to be President of the United States, and I will be glad when I get through; then Tad and I are going out to Springfield, Illinois. I have bought a farm out there and I don't care if I again earn only twenty-five cents a day. Tad has a mule team, and we are going to plant onions.''

Then he asked me, ``Were you brought up on a farm?'' I said, ``Yes; in the Berkshire Hills of Massachusetts.'' He then threw his leg over the corner of the big chair and said, ``I have heard many a time, ever since I was young, that up there in those hills you have to sharpen the noses of the sheep in order to get down to the grass between the rocks.'' He was so familiar, so everyday, so farmer-like, that I felt right at home with him at once.

He then took hold of another roll of paper, and looked up at me and said, ``Good morning.'' I took the hint then and got up and went out. After I had gotten out I could not realize I had seen the President of the United States at all. But a few days later, when still in the city, I saw the crowd pass through the East Room by the coffin of Abraham Lincoln, and when I looked at the upturned face of the murdered President I felt then that the man I had seen such a short time before, who, so simple a man, so plain a man, was one of the greatest men that God ever raised up to lead a nation on to ultimate liberty. Yet he was only ``Old Abe" to his neighbors. When they had the second funeral, I was invited among others, and went out to see that same coffin put back in the tomb at Springfield. Around the tomb stood Lincoln's old neighbors, to whom he was just ``Old Abe." Of course that is all they would say.

Did you ever see a man who struts around altogether too large to notice an ordinary working mechanic? Do you think he is great? He is nothing but a puffed-up balloon, held down by his big feet. There is no greatness there.

Who are the great men and women? My attention was called the other day to the history of a very little thing that made the fortune of a very poor man. It was an awful thing, and yet because of that experience he--not a great inventor or genius--invented the pin that now is called the safety-pin, and out of that safety-pin made the fortune of one of the great aristocratic families of this nation.

A poor man in Massachusetts who had worked in the nail-works was injured at thirty-eight, and he could earn but little money. He was employed in the office to rub out the marks on the bills made by pencil memorandums, and he used a rubber until his hand grew tired. He then tied a piece of rubber on the end of a stick and worked it like a plane. His little girl came and said, ``Why, you have a patent, haven't you?'' The father said afterward, ``My daughter told me when I took that stick and put the rubber on the end that there was a patent, and that was the first thought of that.'' He went to Boston and applied for his patent, and every one of you that has a rubber-tipped pencil in your pocket is now paying tribute to the millionaire. No capital, not a penny did he invest in it. All was income, all the way up into the millions.

But let me hasten to one other greater thought. ``Show me the great men and women who live in Philadelphia.'' A gentleman over there will get up and say: ``We don't have any great men in Philadelphia. They don't live here. They live away off in Rome or St. Petersburg or London or Manayunk, or anywhere else but here in our town.'' I have come now to the apex of my thought. I have come now to the heart of the whole matter and to the center of my struggle: Why isn't Philadelphia a greater city in its greater wealth? Why does New York excel Philadelphia? People say, ``Because of her harbor.'' Why do many other cities of the United States get ahead of Philadelphia now? There is only one answer, and that is because our own people talk down their own city. If there ever was a community on earth that has to be forced ahead, it is the city of Philadelphia. If we are to have a boulevard, talk it down; if we are going to have better schools, talk them down; if you wish to have wise legislation, talk it down; talk all the proposed improvements down. That is the only great wrong that I can lay at the feet of the magnificent Philadelphia that has been so universally kind to me. I say it is time we turn around in our city and begin to talk up the things that are in our city, and begin to set them before the world as the people of Chicago, New York, St. Louis, and San Francisco do. Oh, if we only could get that spirit out among our people, that we can do things in Philadelphia and do them well!

Arise, ye millions of Philadelphians, trust in God and man, and believe in the great opportunities that are right here not over in New York or Boston, but here--for business, for everything that is worth living for on earth. There was never an opportunity greater. Let us talk up our own city.

But there are two other young men here to- night, and that is all I will venture to say, because it is too late. One over there gets up and says, ``There is going to be a great man in Philadelphia, but never was one.'' ``Oh, is that so? When are you going to be great?'' ``When I am elected to some political office.'' Young man, won't you learn a lesson in the primer of politics that it is a *prima facie* evidence of littleness to hold office under our form of government? Great men get into office sometimes, but what this country needs is men that will do what we tell them to do. This nation--where the people rule--is governed by the people, for the people, and so long as it is, then the office-holder is but the servant of the people, and the Bible says the servant cannot be greater than the master. The Bible says, ``He that is sent cannot be greater than Him who sent Him.'' The people rule, or should rule, and if they do, we do not need the greater men in office. If the great men in America took our offices, we would change to an empire in the next ten years.

I know of a great many young women, now that woman's suffrage is coming, who say, ``I am going to be President of the United States some day.'' I believe in woman's suffrage, and there is no doubt but what it is coming, and I am getting out of the way, anyhow. I may want an office by and by myself; but if the ambition for an office influences the women in their desire to vote, I want to say right here what I say to the young men, that if you only get the privilege of casting one vote, you don't get anything that is worth while. Unless you can control more than one vote, you will be unknown, and your influence so dissipated as practically not to be felt. This country is not run by votes. Do you think it is? It is governed by influence. It is governed by the ambitions and the enterprises which control votes. The young woman that thinks she is going to vote for the sake of holding an office is making an awful blunder.

That other young man gets up and says, ``There are going to be great men in this country and in Philadelphia.'' ``Is that so? When?'' ``When there comes a great war, when we get into difficulty through watchful waiting in Mexico; when we get into war with England over some frivolous deed, or with Japan or China or New Jersey or some distant country. Then I will march up to the cannon's mouth; I will sweep up among the glistening bayonets; I will leap into the arena and tear down the flag and bear it away in triumph. I will come home with stars on my shoulder, and hold every office in the gift of the nation, and I will be great.'' No, you won't. You think you are going to be made great by an office, but remember that if you are not great before you get the office, you won't be great when you secure it. It will only be a burlesque in that shape.

We had a Peace Jubilee here after the Spanish War. Out West they don't believe this, because they said, ``Philadelphia would not have heard of any Spanish War until fifty years hence.'' Some of you saw the procession go up Broad Street. I was away, but the family wrote to me that the tally-ho coach with Lieutenant Hobson upon it stopped right at the front door and the people shouted, ``Hurrah for Hobson!'' and if I had been there I would have yelled too, because he deserves much more of his country than he has ever received. But suppose I go into school and say, ``Who sunk the *Merrimac* at Santiago?" and if the boys answer me, ``Hobson," they will tell me seven-eighths of a lie. There were seven other heroes on that steamer, and they, by virtue of their position, were continually exposed to the Spanish fire, while Hobson, as an officer, might reasonably be behind the smoke-stack. You have gathered in this house your most intelligent people, and yet, perhaps, not one here can name the other seven men.

We ought not to so teach history. We ought to teach that, however humble a man's station may be, if he does his full duty in that place he is just as much entitled to the American people's honor as is the king upon his throne. But we do not so teach. We are now teaching everywhere that the generals do all the fighting.

I remember that, after the war, I went down to see General Robert E. Lee, that magnificent Christian gentleman of whom both North and South are now proud as one of our great Americans. The general told me about his servant, ``Rastus," who was an enlisted colored soldier. He called him in one day to make fun of him, and said, ``Rastus, I hear that all the rest of your company are killed, and why are you not killed?" Rastus winked at him and said, `` 'Cause when there is any fightin' goin' on I stay back with the generals."

I remember another illustration. I would leave it out but for the fact that when you go to the library to read this lecture, you will find this has been printed in it for twenty-five years. I shut my eyes--shut them close--and lo! I see the faces of my youth. Yes, they sometimes say to me, ``Your hair is not white; you are working night and day without seeming ever to stop; you can't be old.'' But when I shut my eyes, like any other man of my years, oh, then come trooping back the faces of the loved and lost of long ago, and I know, whatever men may say, it is evening-time.

I shut my eyes now and look back to my native town in Massachusetts, and I see the cattle-show ground on the mountain-top; I can see the horse- sheds there. I can see the Congregational church; see the town hall and mountaineers' cottages; see a great assembly of people turning out, dressed resplendently, and I can see flags flying and handkerchiefs waving and hear bands playing. I can see that company of soldiers that had re-enlisted marching up on that cattle-show ground. I was but a boy, but I was captain of that company and puffed out with pride. A cambric needle would have burst me all to pieces. Then I thought it was the greatest event that ever came to man on earth. If you have ever thought you would like to be a king or queen, you go and be received by the mayor.

The bands played, and all the people turned out to receive us. I marched up that Common so proud at the head of my troops, and we turned down into the town hall. Then they seated my soldiers down the center aisle and I sat down on the front seat. A great assembly of people a hundred or two--came in to fill the town hall, so that they stood up all around. Then the town officers came in and formed a half-circle. The mayor of the town sat in the middle of the platform. He was a man who had never held office before; but he was a good man, and his friends have told me that I might use this without giving them offense. He was a good man, but he thought an office made a man great. He came up and took his seat, adjusted his powerful spectacles, and looked around, when he suddenly spied me sitting there on the front seat. He came right forward on the platform and invited me up to sit with the town officers. No town officer ever took any notice of me before I went to war, except to advise the teacher to thrash me, and now I was invited up on the stand with the town officers. Oh my! the town mayor was then the emperor, the king of our day and our time. As I came up on the platform they gave me a chair about this far, I would say, from the front.

When I had got seated, the chairman of the Selectmen arose and came forward to the table, and we all supposed he would introduce the Congregational minister, who was the only orator in town, and that he would give the oration to the returning soldiers. But, friends, you should have seen the surprise which ran over the audience when they discovered that the old fellow was going to deliver that speech himself. He had never made a speech in his life, but he fell into the same error that hundreds of other men have fallen into. It seems so strange that a man won't learn he must speak his piece as a boy if he in- tends to be an orator when he is grown, but he seems to think all he has to do is to hold an office to be a great orator.

So he came up to the front, and brought with him a speech which he had learned by heart walking up and down the pasture, where he had frightened the cattle. He brought the manuscript with him and spread it out on the table so as to be sure he might see it. He adjusted his spectacles and leaned over it for a moment and marched back on that platform, and then came forward like this-- tramp, tramp, tramp. He must have studied the subject a great deal, when you come to think of it, because he assumed an ``elocutionary'' attitude. He rested heavily upon his left heel, threw back his shoulders, slightly advanced the right foot, opened the organs of speech, and advanced his right foot at an angle of forty- five. As he stood in that elocutionary attitude, friends, this is just the way that speech went. Some people say to me, ``Don't you exaggerate?'' That would be impossible. But I am here for the lesson and not for the story, and this is the way it went:

``Fellow-citizens--'' As soon as he heard his voice his fingers began to go like that, his knees began to shake, and then he trembled all over. He choked and swallowed and came around to the table to look at the manuscript. Then he gathered himself up with clenched fists and came back: ``Fellow-citizens, we are Fellow-citizens, we are--we are--we are--we are--we are--we are very happy--we are very happy--we are very happy. We are very happy to welcome back to their native town these soldiers who have fought and bled--and come back again to their native town. We are especially--we are especially--we are especially. We are especially pleased to see with us to-day this young hero'' (that meant me)--``this young hero who in imagination'' (friends, remember he said that; if he had not said ``in imagination'' I would not be egotistic enough to refer to it at all)--``this young hero who in imagination we have seen leading--we have seen leading--leading. We have seen leading his troops on to the deadly breach. We have seen his shining--we have seen his shining--his shining--his shining sword--flashing. Flashing in the sunlight, as he shouted to his troops, `Come on'!''

Oh dear, dear, dear! how little that good man knew about war. If he had known anything about war at all he ought to have known what any of my G. A. R. comrades here to-night will tell you is true, that it is next to a crime for an officer of infantry ever in time of danger to go ahead of his men. ``I, with my shining sword flashing in the sunlight, shouting to my troops, `Come on'!'' I never did it. Do you suppose I would get in front of my men to be shot in front by the enemy and in the back by my own men? That

86

is no place for an officer. The place for the officer in actual battle is behind the line. How often, as a staff officer, I rode down the line, when our men were suddenly called to the line of battle, and the Rebel yells were coming out of the woods, and shouted: ``Officers to the rear! Officers to the rear!" Then every officer gets behind the line of private soldiers, and the higher the officer's rank the farther behind he goes. Not because he is any the less brave, but because the laws of war require that. And yet he shouted, ``I, with my shining sword--" In that house there sat the company of my soldiers who had carried that boy across the Carolina rivers that he might not wet his feet. Some of them had gone far out to get a pig or a chicken. Some of them had gone to death under the shell-swept pines in the mountains of Tennessee, yet in the good man's speech they were scarcely known. He did refer to them, but only incidentally. The hero of the hour was this boy. Did the nation owe him anything? No, nothing then and nothing now. Why was he the hero? Simply because that man fell into that same human error--that this boy was great because he was an officer and these were only private soldiers.

Oh, I learned the lesson then that I will never forget so long as the tongue of the bell of time continues to swing for me. Greatness consists not in the holding of some future office, but really consists in doing great deeds with little means and the accomplishment of vast purposes from the private ranks of life.

To be great at all one must be great here, now, in Philadelphia. He who can give to this city better streets and better sidewalks, better schools and more colleges, more happiness and more civilization, more of God, he will be great anywhere. Let every man or woman here, if you never hear me again, remember this, that if you wish to be great at all, you must begin where you are and what you are, in Philadelphia, now. He that can give to his city any blessing, he who can be a good citizen while he lives here, he that can make better homes, he that can be a blessing whether he works in the shop or sits behind the counter or keeps house, whatever be his life, he who would be great anywhere must first be great in his own Philadelphia.